Caterpillar counting

I can count up to 10 objects

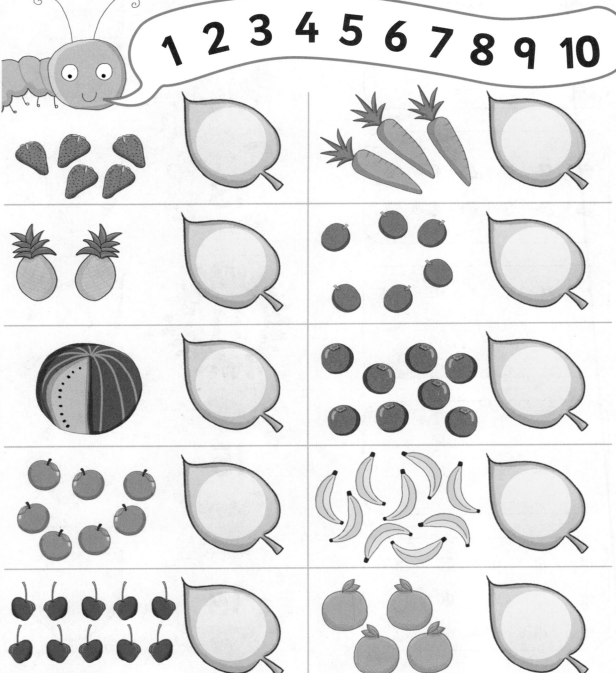

Teacher's notes

The caterpillar is hungry. Look at each set in turn and count the number of items that he is going to eat. Write the matching number on the leaf.

3

Date: _____

Counting minibeasts

I can count up to 15 objects

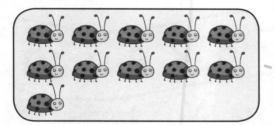

12

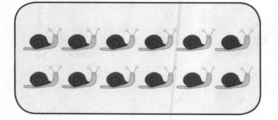

13

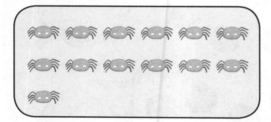

11

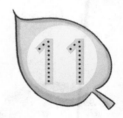
1 1

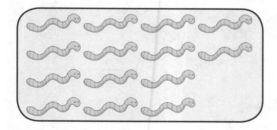

15

14

Teacher's notes

Count the minibeasts shown in each set. Join each group to the flower that shows the correct number. Then write the same number on the leaf next to it.

4

Busy bee counting

I can count up to 20 objects

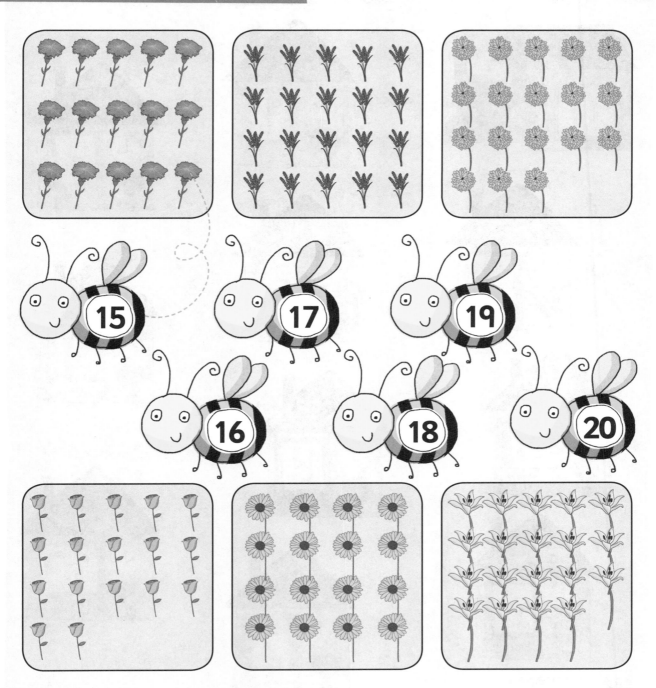

Teacher's notes

To find out which garden each bee visits, count the flowers shown in each garden. Join each garden to the bee that shows this number.

5

Numbers 0 to 10

Date: _____

I can say numbers in order from 0 to 10

Teacher's notes

The train is starting from the engine shed at 0, and will be calling at all of the stations from 1 to 10. Trace the number on each station waiting room to find out the number of each station. Then, draw a line to show which way the train travelled from 0–10.

6

Through the maze

I can say numbers in order from 1 to 20

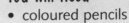

Date: _____

You will need:
• coloured pencils

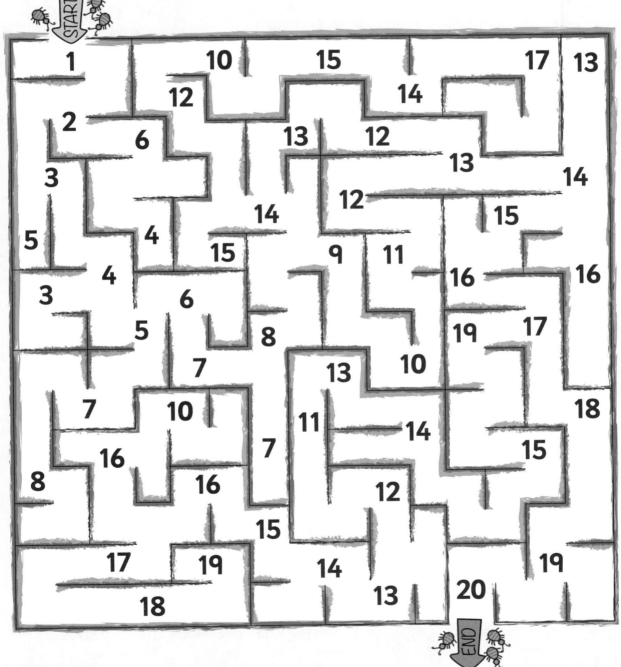

Teacher's notes

Start at the number 1. Use the sequence of numbers 1 to 20 to help find the path through the maze. Colour the route taken.

Bug spot counting

Date: _____

I can find 1 more and 1 less than a number from 1 to 10

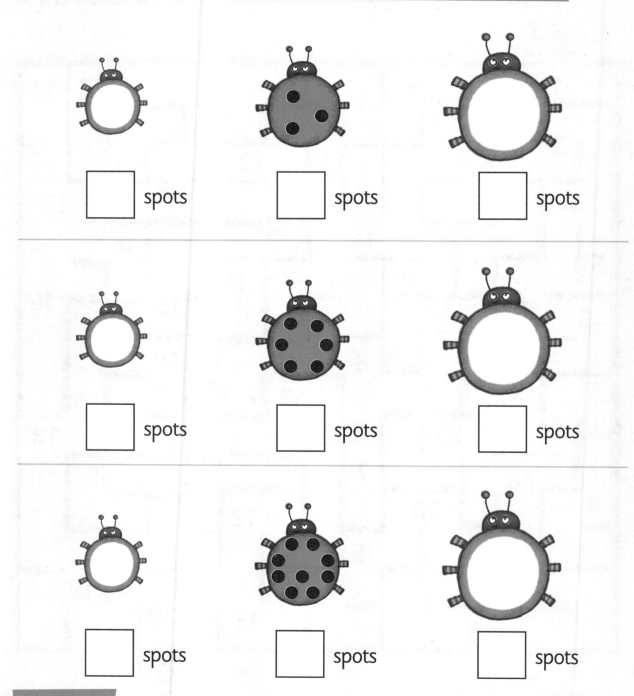

☐ spots ☐ spots ☐ spots

☐ spots ☐ spots ☐ spots

☐ spots ☐ spots ☐ spots

Teacher's notes

In each row, count the spots on the middle bug. Then draw one less number of spots on the left bug, and one more on the right bug. Write the matching number underneath each bug.

Pirate treasure (more)

I can find 1 more than a number from 10 to 20

Teacher's notes

Look at the number on each ship. Match the ship with the island that shows **one more** than this number.

Pirate treasure (less)

I can find 1 less than a number from 10 to 20

Teacher's notes

Look at the number on each pirate. Match it to the treasure chest showing the number that is **one less**.

Date: _____

Adding birds

I can find how many altogether by adding

| How many birds are there altogether? |

| How many birds are there altogether? |

| How many birds are there altogether? |

| How many birds are there altogether? |

Teacher's notes

How many birds are sitting on the first wall? And how many are flying down? How many are there altogther? Write the answer in the space. Now work out the total number of birds in each of the other three pictures and write the answers in the spaces.

Maths magician addition

I can add two groups of objects and say how many there are altogether

You will need:
- coloured pencils

 and  makes  altogether

 and  makes  altogether

 and  makes  altogether

 and  makes  altogether

Teacher's notes

Look at the ingredients in the first pair of jars. Count them and draw that number of ingredients in the cauldron. Then write the total in the box. Do the same for the other pairs of jars.

12

Date: _____

Subtracting cats

I can find how many are left by subtracting

How many cats are left?

How many cats are left?

How many cats are left?

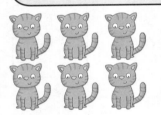 

How many cats are left?

Teacher's notes

How many cats are there in the first picture? How many are walking away? How many are left? Write the answer in the space. Now work out how many cats are left in each of the other three sets of pictures and write the answers in the spaces.

Date: _____

Disappearing spells

I can take some objects away from
a group and say how many are left

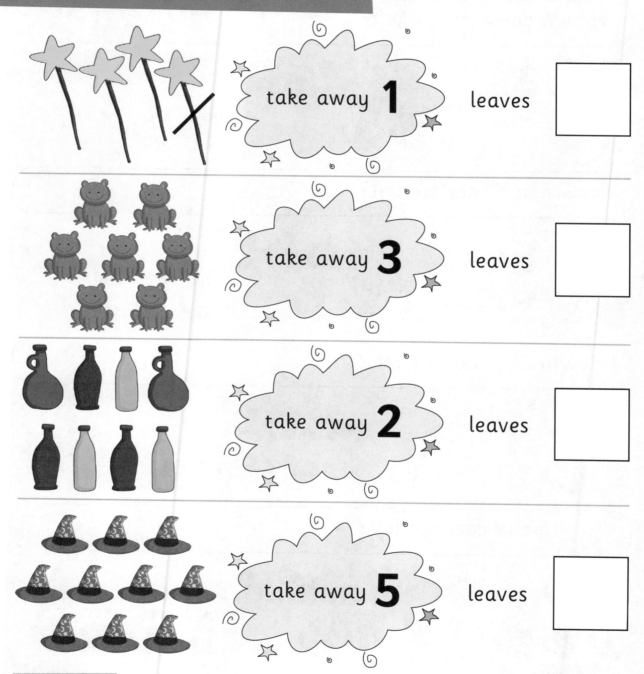

take away **1** leaves

take away **3** leaves

take away **2** leaves

take away **5** leaves

Teacher's notes

Look at each set of objects in turn and look at the number to take away (disappear). Cross out the
number of objects to disappear and then count how many are left. Write this number in the box.

Date: _____

Ladybird spots

I can double and halve numbers

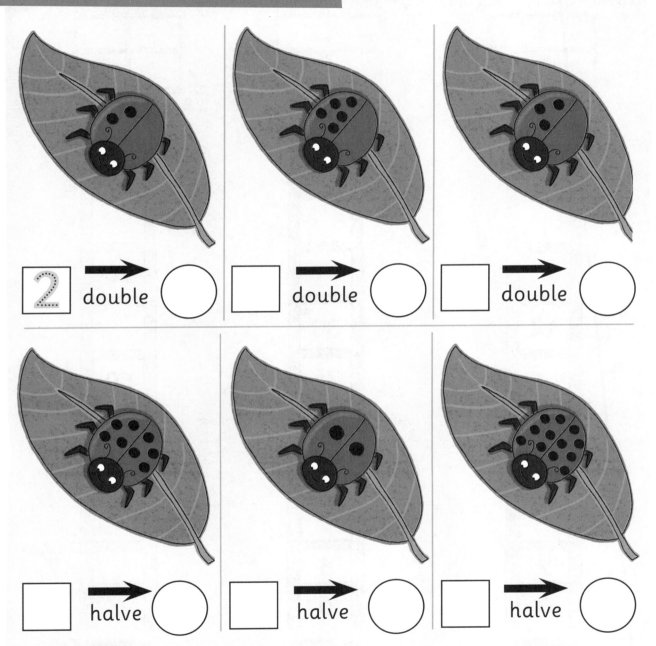

| 2 | double ◯ | | double ◯ | | double ◯ |

| | halve ◯ | | halve ◯ | | halve ◯ |

Step counting

I can count in 2s, 5s and 10s

Date: _____

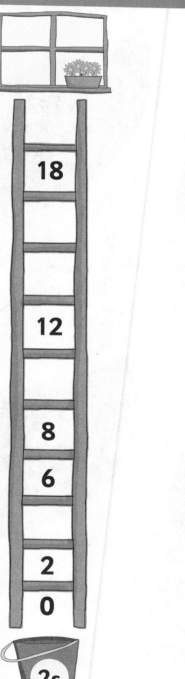

2s ladder	5s ladder	10s ladder
18	45	90
		70
12	30	50
	25	40
8		30
6	15	
2	5	
0	0	0

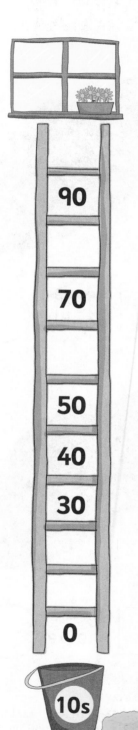

2s

5s

10s

Teacher's notes

Fill in the missing numbers by counting in 2s, 5s or 10s.

Monster socks

I can solve problems involving grouping

Date: _____

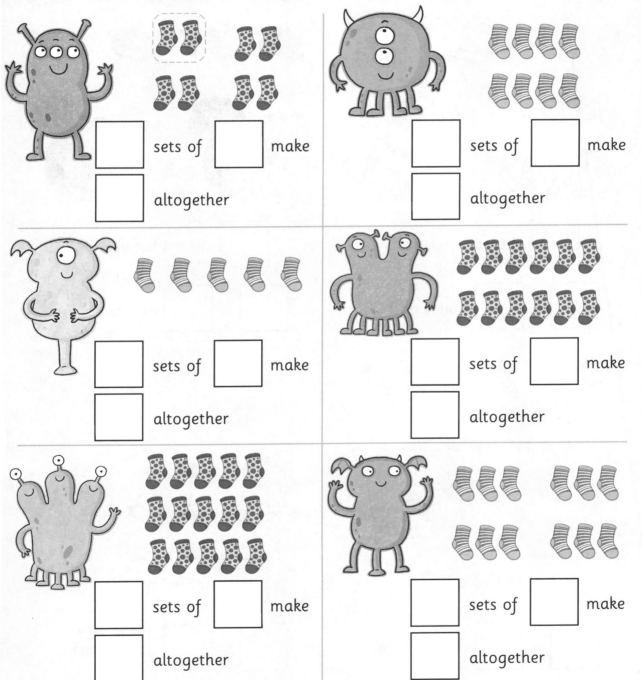

sets of [ ] make [ ]

altogether [ ]

sets of [ ] make [ ]

altogether [ ]

sets of [ ] make [ ]

altogether [ ]

sets of [ ] make [ ]

altogether [ ]

sets of [ ] make [ ]

altogether [ ]

sets of [ ] make [ ]

altogether [ ]

Teacher's notes

Look at each monster. Group the socks according to the number of legs the monster has. Count the number of sets to work out the number of socks in total. Write the number of socks in the box underneath.

Fair sharing

Date: _____

I can solve problems involving sharing

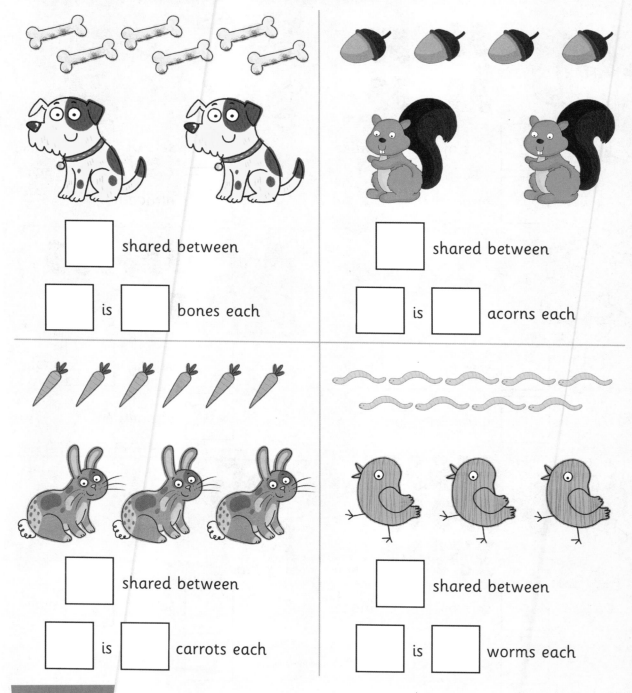

shared between

☐ is ☐ bones each

shared between

☐ is ☐ acorns each

shared between

☐ is ☐ carrots each

shared between

☐ is ☐ worms each

Date: _____

Flat shapes

I can recognise 2D shapes

You will need:
• coloured pencils

| circle | square | triangle | rectangle |

Teacher's notes

Look at the shapes in turn and decide if it's a circle, square, triangle or rectangle. Draw a line from the shape in the corresponding box.

Shape sizes

Date: _____

I can talk about the shape and size of different solid and flat shapes

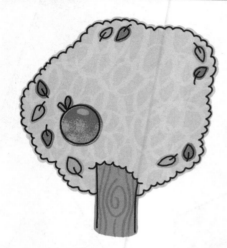

Draw a collection of **smaller** circles

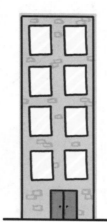

Draw a **taller** rectangle

Draw a **bigger** square

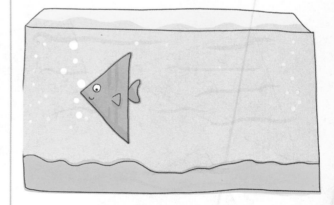

Draw a collection of **smaller** triangles

20

Date: _____

Solid shapes

I can recognise solid shapes in the world around me

Cylinder	Cube	Cone	Sphere	Cuboid

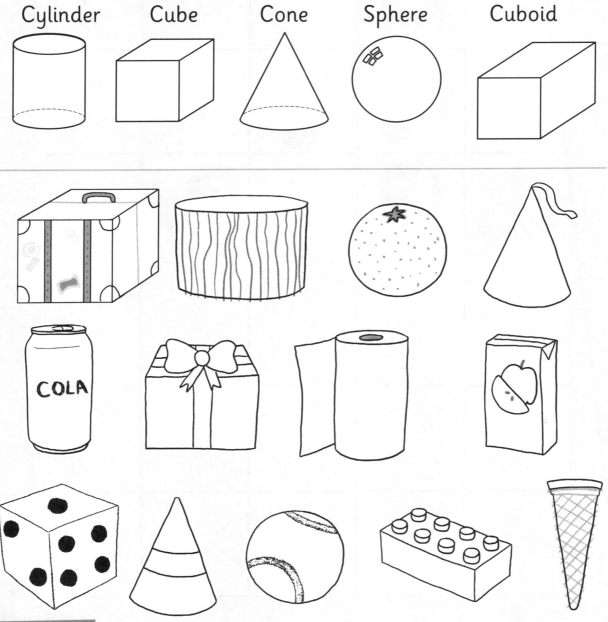

COLA

Teacher's notes

Look at the five 3D shapes at the top of the page. Colour each one a different colour. Then colour each of the 3D objects below to match the corresponding shape.

Position pictures

I can talk about where things are

Date: _____

Teacher's notes

Look at the pictures in the top grid. Draw pictures in the grid underneath so that it looks exactly the same. Tell your helper where you have put each picture. Try to use the words 'next to', 'underneath', 'above' and any other position words you know.

Date: _____

I can compare distances

long   short

long   short

long   short

long   short

Teacher's notes

Discuss each picture with the child. The child circles 'long' or 'short' to describe the distance that each picture shows.

23

Date: _____

Sizes

I can compare objects using the words **longer**, **shorter**, **bigger** and **smaller**

You will need:
• coloured pencils

Draw a **longer** scarf.

Draw a **shorter** pencil.

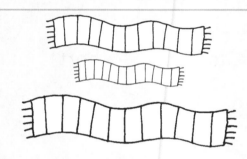

Colour the **shortest**.

Colour the **longest**.

smaller

bigger

bigger

smaller

Teacher's notes

Top section: Follow the instructions to draw a longer or shorter object in each set, and to colour the longest or shortest object in each set.
Bottom section: Draw a bigger and smaller object in each row.

24

Heavier and lighter

I can compare two objects and say
which is heavier and which is lighter

Date: _____

You will need:
- a balance
- items suitable for weighing
- coloured pencils

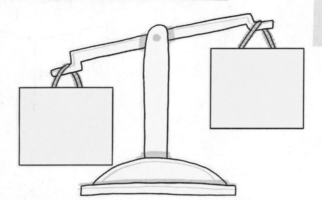

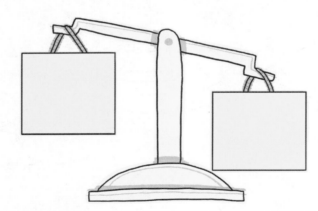

Teacher's notes

Using a balance, and choosing from classroom resources, investigate weight. Find items that are
heavier and lighter than each other. Draw these items on the pans of the balances to show which were
heavier and lighter. Tell your helper what you found and ask them to write a short sentence under each
balance.

25

Date: _____

Capacity colouring

I can compare objects and say which holds more or less

You will need:
- red and blue coloured pencils

Teacher's notes

Look at each set of objects. In each set, colour in red the object that holds the most. Colour in blue the object that holds the least.

Time is ticking

Date: _____

I can use everyday words to talk about time

You will need:
- coloured pencils

Draw something you did yesterday.

Draw something you did this morning.

Draw something that takes a long time to do.

Draw something that is quick to do.

Teacher's notes

Discuss each of the statements with the child. In discussion with the child, decide on the most suitable event to draw in each box.

Date: _____

Currant buns

I can solve money problems

Teacher's notes

Currant buns cost 1p each. Count the pennies that each child has, and colour the currant buns they are able to buy with this amount. Write the total in the circle above each of the coins that have the equivalent value.

28

Date: _____

Pocket money purses

I can solve money problems

 is the same as

p

 is the same as

p

 is the same as

p

 is the same as

p

 is the same as

p

Teacher's notes

Look at the coins in each purse. Colour the pennies next to each purse to show the same amount.
Write this amount below the child to show how much pocket money each child has.

Birthday patterns

I can copy and continue simple patterns

Date: _____

You will need:
• coloured pencils

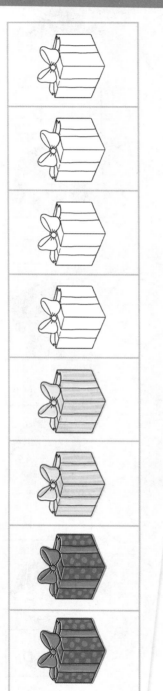
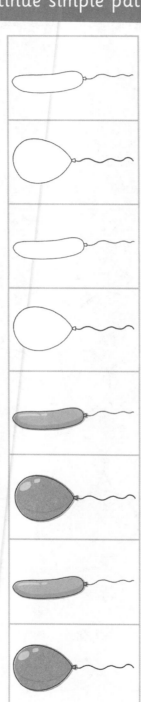
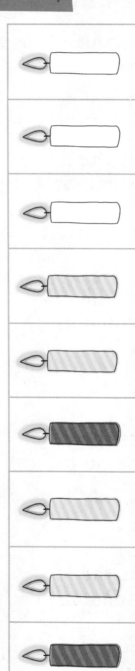
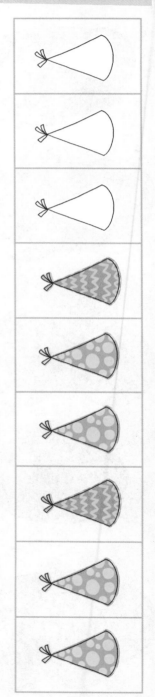

Teacher's notes

Look at each pattern in turn. Complete each one by colouring the objects in each row.